AF824480

Foundations

for Great Enterprise and True Wealth

by
Sam Frowine

Institute for
Business Owners

Copyright 2000, 2003, 2008, Sam Frowine

All rights reserved

No part of this book may be reproduced, stored in a retrieval system, or transmitted by any means without the written permission of the author.

Center for Great Enterprise
9115 Harris Corners Parkway
Suite 330
Charlotte, NC 28269
704-597-5156

Cover photograph by Ian Britton
Frontispiece photograph courtesy of PDPhoto.org

The Four Foundations

Perspective
Integrity
Respect
Stewardship

Integrity

- Honesty
- Beliefs Aligned with Actions

Perspective

- Strategic Planning
- Success Factors
- Leadership
- Vision
- Principles of the Living Enterprise

Respect

- *God*
- *People*
- *Time*
- *Money*

Stewardship

- Resource Management
- Succession Planning

Dedication

This book is dedicated to the first business owner who molded my thinking about the value of outcomes that go far beyond money or power or material acquisition: My father, Samuel E. Frowine, Jr., Attorney at Law.

Author Biography

Working with business owners to achieve their goals in life is what drives Sam Frowine. For more than 20 years, he has been owner, CEO, managing partner or investor in more than a dozen enterprises—several very successful and a few not so successful. Sam is also the author of ***Lifeblood: How Successful Business Owners Achieve Wealth***; the ***Business Wealth*** column in ***The Charlotte Business Journal***; and ***Blueprint for Building Great Enterprise.*** He is also creator of the Enterprise Builder System™ and the Business Wealth System™, a methodology for optimizing ownership wealth. He is founder and owner of The Performance Group, Performance Capital Group and the Institute for Business Owners. The focus of Sam's work is coaching entrepreneurs whose companies are the foundation for their portfolio wealth. Sam, whose roots are in Portsmouth, Ohio, holds a doctorate degree in developmental psychology from the University of Cincinnati, where he was a faculty member for seven years. He received his undergraduate and Master of Science degrees from Ohio State University.

Acknowledgements

Everything I've written has grown out of the privilege of sharing life with business owners on their journeys. Any true wisdom in my work is the result of Divine guidance and inspiration.

The opportunity to apply the principles of my work has been given to me as a ground-floor partner in building in Lake Forest Church, where we are being provided with immeasurably more than we could begin to imagine. The four founding elders from whom I have learned much are Jack Cathey, Rusty Andrews, Dave Grimes and Mike Moses.

My wife, Velda, is the prayer warrior who intervenes on my behalf every day when I walk out into the world of business ownership.

I am grateful to be an instrument for providing business owners with a model for applying deeply-held principles to their role as business owners. The very challenges and triumphs of business ownership offer every owner the opportunity to live these principles more deeply and to experience what it is to be provided with immeasurably more than they have imagined.

- Sam Frowine

Table of Contents

True Wealth

The state of mind and spirit
that occurs when people
use their God-given talent
and their passion
to produce substantial
financial return, while
gaining deep satisfaction
from work that is significant.

Chapter One
Creating Opportunities for True Wealth

"True Wealth marks the dividing line between those who keep looking for satisfaction from their money and those who understand that money merely offers an opportunity to seek out sources of significance and deep reward."

What is the difference between a business owner who has made a lot of money and one who has financial well-being? One who longs to look down from the top of the heap and one who arrives at the mountaintop? One who has managed to acquire the trappings of wealth and one who enjoys financial, personal and professional peace of mind? One who asks, "Is this all there is?" and one who has built an asset of perpetuity and makes a valuable contribution to the world.

The difference is True Wealth™.

When people talk about wealth, what most of us mean is "rich." And "rich" is about as far from "wealth" as we can get. In fact, some people who ***look*** rich sometimes have actually acquired a lot of stuff and a lot of debt, but possess little of true value. Chances are that's not what you want out of life.

What you want is True Wealth.

Having It All and Wanting More

Business owners work hard—grindingly hard—year after year to achieve their dreams. Sometimes they miss the mark and give up. Sometimes they fall short and keep going. Sometimes they get everything they've dreamed of and still wonder why they feel as if something is missing.

> Owners have the opportunity to seek out a purpose greater than the work they're doing or the things they're accumulating.

They've got the house at the lake, the three-car garage, the boat, the country club membership…the list goes on. Of course, some of them also have debt that goes on forever, as well. Some of them may have a spouse who suffered every setback and every sacrifice with them over the years. Maybe they have kids that make them proud. A little recognition in the community, a little influence in their industry…they've made it. They've arrived.

But the feeling that something is missing just doesn't go away. Everything they feel they've accomplished feels like vapor—there's nothing permanent they can hold onto.

They start to wonder if they're the only ones who feel that way. They wonder if there's something more to business—to life—than the next deal, the next financial milepost, the next high-ticket toy. They've worked hard, taken calculated risks and achieved success by many of the world's standards.

So why aren't they satisfied?

They've been looking for their satisfaction in the wrong place. They are business owners at a crossroad—the perhaps the most significant crossroad they'll ever face.

Choices at the Crossroad

The journey of business ownership leads to a number of crossroads. We'll talk a little about each of them in Chapter 4. But this particular crossroad is more than the juncture in the highway. This crossroad is the spaghetti junction of all the crossroads. Unless you navigate this one successfully, the others may lead to a series of dead ends.

These owners who know they're missing something vital are ready for the journey to True Wealth.

I've had the privilege of meeting many business owners when they reach their personal crossroad. In the work of my consulting company, my partners and associates and I sometimes see the wreckage—personal, professional—of trying to ignore the crossroad. They see the look in business owners' eyes that says, ***Is this all there is? Is this all I have to show for nearly killing myself to keep this business alive?***

The great gift in this painful juncture, of course, is that the crossroads offer a chance to head in a different direction. At the crossroads, owners have the opportunity to seek out a purpose greater than the work they're doing or the things they're accumulating.

Some engage in the journey. Some choose the comfort of continuing in the same direction.

Money Plus Meaning

What is True Wealth? Here's what I believe: True Wealth is about more—***much more***—than the balance

---

A True True-Wealth Story

Early in my career in developmental psychology, I worked in prisons, with young children growing up in poverty, with Welfare mothers who couldn't see their way to a better life. I've always had a heart for people in hard circumstances.

One of the reasons I ended up in business was the frustration of finding that often the best I could do was Band-Aid these overwhelming problems. This ultimately lead me to my calling of growing great leaders on the platform of private enterprise, in order to make a real and permanent difference in the world.

Recently, that early experience came back to me when our church opened its doors for homeless women once a month. They slept in our vestibule and showered in our rest rooms and ate food our members brought in. I sat with them and talked with them. Something was ignited in me.

I made a commitment. Every Friday morning, I would show up at the Hope Café on the edges of downtown, where the sheet, the bank statement, the material things accumulated. True Wealth goes much deeper.

I'm a business man, so you know I'm not saying for a single minute that True Wealth has nothing to do with money. True Wealth has a place for money, an important place. But True Wealth marks the dividing line between those who keep looking for satisfaction from their money and those who understand that money merely offers an opportunity to seek out sources of significance and deep reward.

True Wealth means something different for each of us. Certainly many of us begin by believing wealth is about our

---

homeless show up to get in out of the cold or the rain or the heat to receive a cup of morning coffee. Week after week, I showed up with a box of 75 hot biscuits. I looked into the eyes of people I might've passed without even seeing before. I talked with them and listened to them and heard once again the stories of their overwhelming challenges.

My box grew to 90 biscuits. A hundred.

Something in me grew, too. Grew so big I knew I wanted to do more than show up once a week with biscuits.

I talked with the leaders of NeXus, the church that sponsors Hope Café, about putting my resources to work building a program for connecting business owners and community leaders with the homeless who are also often jobless.

My experiment in Social Entrepreneurship ultimately gave me something worth far more than any other investment I could've made with my time and money. It added to the balance in my True Wealth account.

---

deeply-held desire for freedom to make our own choices and control our own destiny.

But another common thread for the classic entrepreneur is the desire to achieve a level of financial success that enables them to impact the world and leave behind a legacy. Not just an inheritance, but a legacy. We want to build something of lasting value. Something that improves lives and changes the world for the better in some way large or small.

We've all seen buildings named after someone who achieved something great, or at least had the money to bring together a lot of bricks and mortar. That's one kind of legacy. You might hear about another kind of legacy when you listen to a public radio station that's receiving support from a foundation named after a person or a family. That's another kind of legacy. We all know what kind of legacy Mark Twain or Charles Dickens left—well-constructed words often echo for generations. That's part of the legacy I hope to leave—a body of knowledge that's written down for future generations to use and benefit from.

> The money isn't just showing up as a vacation house on the ocean; it's serving to fill an owner's life with meaning by supporting positive change on a larger scale.

Legacies come in a lot of packages.

Other business owners move through the Start-Up and survival stages and mature into the Enterprise Asset and Wealth Builder stages, where they inevitably begin to ask themselves, "What's next?" These owners may become part of the Social Entrepreneurship movement to address issues of injustice or poverty around the world. I've worked with dozens of business owner who have used profits to sponsor mission trips around the world, or who work diligently and selflessly to provide economic opportunities for the homeless in their own back yard. One woman's long-range vision is to funnel her profits into a residency and training program for women who desperately need a second chance.

> Legacies come in a lot of packages.

All these people have achieved one of the markers of True Wealth. They've arrived at a level of financial reward that not only supports a business, it makes the world a better place. The money isn't just showing up as a vacation house on the ocean; it's serving to fill an owner's life with meaning by supporting positive change on a larger scale.

Momentum and Opportunity

There's more. True Wealth goes deep.

It undoubtedly touches friends and family. And when a business owner is on the True Wealth journey, the implications extend throughout the business. It typically becomes a transforming experience for the people within the business…which creates positive impact on their families and friends…it's a positive domino effect the full impact of which can't even be measured.

That's because True Wealth is about the integrity with which we conduct our lives and our businesses. It is about the wise perspective that comes with knowledge of and reflection on the work we do. It's about respect for the responsibilities and resources we've been given, and our willingness to be good stewards of those resources.

Integrity. Perspective. Respect. Stewardship.

Those attributes are the Four Foundations™ on which Great Enterprise and True Wealth are built.

> ...ownership becomes conscious and intentional about making the choices that can lead to positive growth and deep transformation.

We'll talk more about the Four Foundations later, along with the concept of Great Enterprise, the laws for building it and the practices of high-performance companies. You'll see how a Great Enterprise differs from a company that's simply making a lot of money or surviving to pay the bills. You'll begin to see how True Wealth principles can become part of the company culture when ownership becomes conscious about making the choices that can lead to positive growth and deep transformation.

Applying True Wealth principles opens the door for people inside the company to understand the premises on which the business is built and to gain trust in ownership and its vision for the future. This allows them to embrace the deeper ***purpose*** of the business, and not just the ***work*** of the business.

This creates momentum for the company as it is transformed into a Great Enterprise™. And a Great Enterprise ***always*** offers opportunities for True Wealth.

Great Enterprise

A Great Enterprise
is always in process. An
organization must remain
committed to continuous
growth and lifelong
learning in order to stay
on the journey to
greatness.

Chapter Two
What is a Great Enterprise?

"A Great Enterprise is identified by the fruits of its actions."

You've heard that business owners who want more out of life than hard work and a house full of stuff may be on the journey to True Wealth.

You've heard that a business must become a Great Enterprise™ before it can hope to provide its owner with True Wealth.

So what exactly ***is*** a Great Enterprise? What does it look like? How is it any different from your average successful business? Does it depend on how many people are employed there, how many locations it operates from, its geographic footprint, how much sales go up from year to year? We've all seen businesses recognized and honored and written about for the kind of measurable growth that could be called "great." Is that where the measurement of a Great Enterprise begins?

As far as the word "enterprise," seeing how different it is from a mom-and-pop or a solo practitioner doesn't take a genius.

So we could define a Great Enterprise in the ways our culture and our economy have taught us. But in my experience, metrics that can be plugged into a spreadsheet are only one part of a Great Enterprise.

> The word "great" implies legendary, noteworthy or of high value. It assumes there is something special or uniquely brilliant about a person, place or thing—in this case, a business. A Great Enterprise can achieve that status when its leaders seek to follow certain principles and practices in making decisions about the business.

Great? Or Not So Great?

Over my years as a business owner, I've walked through countless companies. And I can tell you in less than an hour—in about 15 minutes, in some cases—what the underbelly of the business looks like. I can tell you whether I'm experiencing a Great Enterprise, or a business that's not so great in some aspect or another.

I can tell you whether its people feel proud of what they do. I can tell you how much integrity it has, top to bottom. I can tell you whether people are respected, top to bottom. I can tell you whether growth is healthy, whether people have a sense of the company's future and their role in it, whether people pull the plow together or carve up the field and say, "That's not my job."

As you've probably already figured out, I believe a Great Enterprise is measured by a lot more than traditional success.

I believe a Great Enterprise is identified by the fruits of its actions.

In other words, what's going on in a company? ***Your*** company, for example. What would I see if I walked through your business?

The Fruits of Great Enterprise

Here are some of the fruits a company bears when it's on track to become a Great Enterprise.

- The people of the organization are valued and appreciated by ownership… and they know it.
- The people respect and value their leader.
- The business is financially healthy and prudent with its resources.
- The leaders sense that they are building a legacy that will be passed on to the next generation; they take seriously their role as stewards.
- The business does something brilliantly better than anyone else.
- The company culture centers on people serving people, with customers and clients coming before the desires of special interest groups—you know, owners, leaders and self-serving others.

- Every player understands the metric for which they are responsible as part of winning the game.
- Success is measured on results, not ideas.
- The environment cultivates personal growth and transformation.
- The organization engages in healthy, balanced innovation through continuous adaptation. It ***always*** adapts. We do this through a process we call Think-Execute-Account.
- The people of the business have a passion for their work and understand how their work links to the health of the enterprise.
- The people of the business have a sense of community responsibility, making contributions outside the boundaries of the business.
- The leaders who shape the decisions understand time perspective, beyond the urgent to the balanced. A business is in a marathon, not a sprint.

These are the characteristics of a truly Great Enterprise. Business owners who desire True Wealth encourage this kind of company culture. They are actively engaged in ongoing improvement in these areas. And in doing so, they are ultimately providing their business with a rootedness in the kind of principles that will help it sustain the winds of difficulty every business inevitably faces.

How your company stacks up today is less important than your willingness, as the owner, founder or leader of the organization, to foster a climate for greatness.

The Fruits of Great Enterprise

But how does a company move from wherever it is today to become a Great Enterprise? Can a company turn itself around? How does a new company start out in the right direction from the get-go?

Turning a business around requires great boldness and courage. That's what my company does—helps business owners take the right steps to transform their businesses 90 days at a time. We believe—and we've proven it over and over again—that a company can completely transform itself in twelve 90-day periods in which it commits to change through strategic thinking, excellent execution and continuous learning through evaluation and accountability.

So, yes, most companies—including yours—can begin at any time to operate by Business Wealth principles and launch itself on the journey to becoming a Great Enterprise. You're the owner. It's your choice.

A new company gets a good start in the same way—through a commitment to operating by Business Wealth principles.

It all begins with building on the strongest of foundations.

The Four Foundations

Perspective
Integrity
Respect
Stewardship

CHAPTER THREE
THE FOUR FOUNDATIONS OF GREAT ENTERPRISE

"These Four Foundations of Great Enterprise can also be considered the foundations for True Wealth."

What are the foundational beliefs and values of your business?

That's one of the questions we ask business owners when we're working with them to capture their Grand Vision for the future of their companies. It's always surprising to me how many can't easily answer the question. It's even more surprising how many tell us that their beliefs center around answers that show very little reflection: They believe in good customer service. In building a high-quality product. In taking good care of employees.

Those things are important, of course.

But if you're going to build something substantial, shouldn't the foundation be substantial, too? If you've ever watched the early stages of building a house, you may have seen how deeply the footings are sunk into the ground. The larger the house, the deeper the footings must be in order to support the foundation.

That's what to look for in a Great Enterprise—foundational beliefs that come from somewhere deep and well-grounded.

Cornerstone for Company Culture

As I've studied business owners and their companies over the last two decades, I've identified the foundational truths that under-gird every Great Enterprise. I call them the Four Foundations. And as my understanding of True

---

Uncovering Your Beliefs & Values

- What matters to you most as a business owner?
- When you think about the pain of ownership, what makes it worth walking through the challenges?
- Does your business allow the expression of your most closely held desires and priorities for you life?
- What do you want your business to be known for?
- How will you define the culture upon which your business will build its legacy?
- At the end game, how do you want to be remembered?

Wealth gelled, I came to understand that these Four Foundations of Great Enterprise can also be considered the foundations for True Wealth. They may be expressed differently from company to company; but if it's a Great Enterprise, you'll always find them.

> Perspective is knowledge of the business's what and how, but it also delves deeper to answer the question, "Why?"... The true purpose of the business becomes the rudder that directs the strategies and practices of the business.

The Four Foundations are simple, but deep. And they underpin every principle and practice taught through the Center for Great Enterprise.

These Four Foundations become the cornerstone for the owner's decisions and actions, because the owner sets the tone for the entire enterprise. When the owner is living these principles they become part of the company culture, resulting in opportunities for True Wealth.

And when the owner is only giving lip-service to them, the ground gets shaky under the business.

Abstract Concepts, Concrete Applications

What are the Four Foundations, and what exactly do they mean in the context of a business?

Perspective. Perspective is an abstract idea that shows up in a lot of really practical, concrete ways. It is the guiding principle that lights the way for successfully applying other three Foundations.

Perspective can be boiled down to a single thought that is reflected in countless ways: Perspective is understanding and keeping in mind the Big Picture.

Perspective is the ability to analyze and plan around vision, priorities and money. It is an awareness of the importance of systems and procedures around key elements such as customers and sales and production.

Perspective is knowledge of the business' what and how; but it also delves deeper to answer the question, "Why?"

At its most profound level, Perspective is about an owner's ability to maintain the right relationship to the

---

Uncovering Your Beliefs & Values

We talk a lot about culture building in the world of business today. In private enterprise, the root of culture building begins with the authentic integration of the Four Foundations into every aspect of the business. And authenticity allows for the building of trust.

This becomes the wellspring of the culture you are creating.

These Four Foundations are core to every Great Enterprise. When you live in them, they become part of everything you do in your business.

If you're a linear thinker and all these touchy-feely ideas don't seem to apply to your reality, think of it this way: Your beliefs become the actions you take. Your actions become your habits. Your habits become your character. And your character becomes the DNA of your enterprise.

---

business as it moves through the cycles of change and growth. It is understanding that the business is a living organism, separate and apart from the owner. Perspective is maintaining the healthy boundaries in relationship to this living organism.

And ultimately, Perspective is being aware of the true purpose of the business, which then becomes the rudder that directs the strategies and practices of the business.

Finally, Perspective is the Big Picture that enables owners to survive the inevitable difficulties of the day because they have hope for the future.

> No one, not even the owner, can squander the resources of the company without draining the business of its lifeblood.

With the proper perspective, anything is possible.

Integrity. When we think of integrity, we think of honesty and treating people fairly and keeping everything aboveboard. But at its core, integrity is found in the answer to this question: Do a company's stated beliefs align with its actions?

Integrity exists when the people of the company carry out their daily work in a way that demonstrates their words, actions and beliefs are one and the same.

Does ownership say it wants one thing, but make choices that say otherwise? Here's an example: Ownership says, "I want to take good care of this business so my people will be well-rewarded and have a secure future," then make choices that say, "I'm taking the lion's share for my family so we can afford the luxuries we deserve. After all, I'm the owner. It belongs to me."

Are company messages authentic, not simply words designed to create the right impression? Do company beliefs and practices apply equally across the board?

In today's environment of deception and broken trust, people sometimes believe it is impossible to operate in the world of business with integrity. Here is what I know: To achieve True Wealth, Integrity isn't an option, it's standard operating procedure. To be in Integrity is a discipline of holding ourselves to a higher standard than what is sometimes accepted or expected today.

And like every aspect of the company culture, this starts at the top, with the owner's words, actions and beliefs.

> The Four Foundations play a part in the financial viability of a business—they become a factor in the actual market value of a business.

Respect. Respect is reflected in the owner's attitudes toward the lifeblood resources of the business—the resources of time, money and people that are critical to the health of the business. When the owner demonstrates a lack of respect for these critical resources, so will others throughout the organization.

Respect for the people within the organization creates unity—a spirit of cooperation—and a willingness to innovate, to accept accountability. And as one of the business owners in my circle reminds me, the way we treat our people often becomes the way our people treat our customers or clients.

Respect for both time and money means using those resources on the right priorities to achieve the goals and missions of the business. No one, not even the owner, can squander those resources without draining the business of its lifeblood.

It's also the owner's responsibility to ingrain this same respect in the leaders and employees of the business. This sense of responsibility for protecting the health of the business by respecting its lifeblood resources will become pervasive, part of the collective consciousness in a Great Enterprise.

Stewardship. The resources of a business are not the owner's to hoard or waste. When an owner has been blessed with abundance and success, the owner also bears responsibility for using those resources prudently and well, to create abundance and opportunities for others within the business and the community.

Stewardship is at the heart of the Business Wealth System, which equips owners to prudently calculate the amount of profit which must be extracted from the business in order to build an external base of capital. Growing this capital and putting it to work enables owners to become capitalists to their own and others' enterprises. This is good stewardship of the gifts they've been given.

Market Value, Legacy Value

Perspective. Integrity. Respect. Stewardship.

These four values impact every aspect of a business—its planning, its execution, its financial disciplines, its human capital, its relationship capital, even the quality of its product or service. This is not just "soft" stuff, nice but hardly critical when we're talking about the hard realities of revenue, payroll, profit. Each one of these foundational values play a part in the financial viability of a business—they become a factor in the actual market value of a business.

When they are in place—both within the business and within the owner—it is possible to grow a business with the potential to become a Great Enterprise.

When they are missing, a business not only fails to achieve its full potential, but over time—sometimes quickly, sometimes slowly—the health of an organization begins to diminish.

So it is clear that the Four Foundations also affect the legacy value of a business. They determine what kind of legacy an owner will leave because they are significant determining factors in how an owner will be remembered and what will be carried forward when the owner is no longer in the picture.

The Mind Map at right shows how these Four Foundations permeate the choices and practices of the enterprise, becoming the bedrock of a Great Enterprise.

For most business owners, the decision to build an enterprise on the Four Foundations is one of the most significant crossroads to be faced.

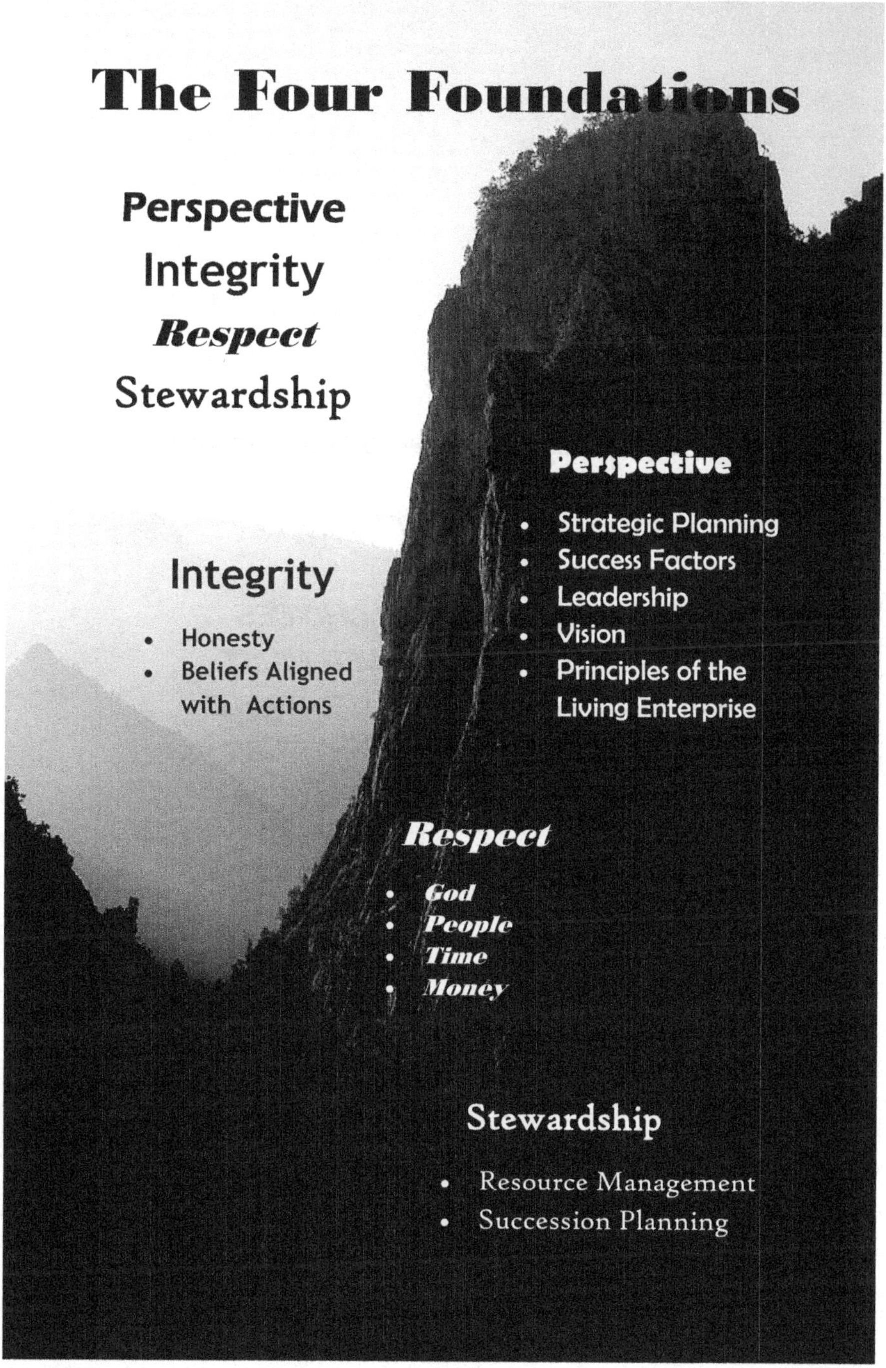

The Four Foundations

Perspective
Integrity
Respect
Stewardship

Perspective

- Strategic Planning
- Success Factors
- Leadership
- Vision
- Principles of the Living Enterprise

Integrity

- Honesty
- Beliefs Aligned with Actions

Respect

- *God*
- *People*
- *Time*
- *Money*

Stewardship

- Resource Management
- Succession Planning

The Big Decision

THE TRANSFORMING CHOICES EVERY BUSINESS OWNER MUST MAKE AT A SERIES OF INEVITABLE JUNCTURES IN THE ROAD OVER THE LIFE OF A BUSINESS.

Chapter Four
The Big Decision: Ownership at the Crossroads

"To avoid stagnation, a business will move through these life stages continuously as it evolves, builds, reinvents itself."

Every business owner faces a crossroads, a juncture in the road when he or she must make a Big Decision™ about the direction of the business. The decision centers on the owner's role in the business and what he or she chooses to allow the enterprise to become.

Avoiding this crossroads isn't an option. A business is a living organism with a life of its own, separate and apart from the owner. It moves through developmental stages that are predictable and similar to the unfolding of any other living organism. To avoid stagnation, a business will move through these life stages continuously as it evolves, builds, reinvents itself.

And reinventing the business inevitably means reinventing ownership. For a business to grow and flourish, the owner's leadership role must continuously evolve. Only by making the decision to change their role in and their relationship to the business are owners able to tap into and develop the best thinking and talents of others in their organizations. This lays the groundwork for an owner-centered enterprise to become a leader-centered enterprise.

Decisions at the Crossroads

Most business owners will face at least one—usually more than one—of these major crossroads. At each crossroad is the opportunity to build wealth. The classic crossroads are:

- Grow: Is it time to expand product/service offering, open additional locations, expand geographically, franchise?
- Re-engineer: Do changes in the marketplace make it imperative for you to make significant changes, or lose market share?
- Capitalize: Does a significant opportunity make this the right time to consider an infusion of outside capital?
- Succeed to the Next Generation: Is it time to begin releasing control to the next generation, either leaders who have grown up within the business or the next generation of family members?
- Sell: Do market or industry conditions make this the right time to sell? Will that enable you to achieve your ownership objectives?
- Liquidate: Is it time to close the doors and salvage what you can?

Choices at the Crossroads

A business is a living organism, with a life of its own and natural life cycles that cannot be avoided. This continuous and inevitable change forces many significant choices for owners at the crossroads.

Should the business grow? Retool in response to market changes? Acquire funds to capitalize a critical move? Sell?

The life stages that bring owners to these crossroads are predictable, must be planned for and clearly have significant impact on the future of a business...and its potential for wealth-building...and its legacy potential.

But few ownership decisions are more important than the decision to build a Great Enterprise as the foundation for True Wealth. It has implications for the owner's role in the business, for the strategies that drive the business, for the kinds of people who will be drawn to the business, and for the end game, when reputations and legacies are made.

Owners at the crossroads will make one of the following Big Decisions:

- *Change by default, in which the owner tries to operate the business just as it is, hoping change won't outpace the original pleasure found in starting the business.*
- *Simplify by scaling down – in other words, to return to the original scope and purpose of the business.*
- *Choose the transformational journey to True Wealth by redefining the owner's role.*

The choice boils down to one thing: Are you prepared to transform yourself?

Conscious Choice

Does anybody really want to change? Growth pains aren't fun. Don't we all, deep down, hope to arrive at some comfortable status quo and settle in for a nice, comfortable ride for the rest of the journey?

Even those of us with an entrepreneurial mindset dream about a day when things aren't so hard, when the changes aren't so tumultuous.

To change or not to change isn't really the choice we're faced with, however. The choice is to make a ***conscious*** choice about change or to have the choice thrust upon us.

I doubt if I even have to tell you which choice, ultimately, is more painful.

This process of choosing True Wealth isn't a single decision. It's a series of decisions about the way you do business, about how you view the business, about what you want from and for the business. And it starts with two of the most basic questions:

- What do you deeply, truly want?
- What are you willing to do to get it?

These are the first two questions on the pathway to True Wealth.

THE BIG DECISION
...Choices at the Crossroads
Growth
Reinvention
Capitalization
Succession
Sell
Liquidate
The ultimate crossroad is whether to engage the journey to True Wealth

The Seven Laws

The Law of Trust
The Law of Customers
The Law of Profit
The Law of Optimization
The Law of Focus
The Law of Harvest
The Law of Balance

Chapter Five
The Seven Laws for Building Great Enterprise

"For optimal performance, the business must be balanced in every dimension of its being."

Certain laws rule the universe. The law of gravity, for example: If you drop something, it falls.

The same is true for building Great Enterprise. Certain uncompromising laws apply in order for a business to become great.

In the case of building Great Enterprise, there are seven. Seven laws that an owner must know, understand and commit to implementing in a company. These Seven Laws of Great Enterprise™ must be in place in an organization before it is capable of creating Business Wealth.

As you read about these laws in this chapter, I'm sure you'll find them to be self-evident and uncomplicated. In fact, I sometimes speak of them as Lessons from the Lemonade Stand. Or the paper route, or whatever fledgling enterprise you might've engaged in as a kid.

1. The Law of Trust

For people to feel free to make good decisions, they must be able to trust in leadership. For people to be released to take the risks required for an organization to grow requires trust. Trust starts at the top, but it moves in both directions. Ownership must demonstrate appropriate levels of trust for the people of the organization and must also inspire trust.

2. The Law of Customers

An enterprise is as healthy as its relationships with customers. Healthy, profitable enteprise growth comes from happy, healthy customers.

3. The Law of Profit

A business without profit is not a business asset. It is a liability. Inherent in the term "business" is the result called profit.

4. The Law of Optimization

To achieve the healthiest enterprise possible, a company must always make the highest and best use of all its resources of time, money and people. Slotting people—

typically our most expensive asset—for their best use and investing to cultivate their talents is an obvious one. Spending time where it will yield the best results is another. Prudent deployment of revenue is another. Making sure our facilities are being fully utilized, our systems are operating at peak efficiency, every asset yielding the best possible return. Here's one we overlook—our own time and talent. If we spend too much time down in the weeds, we're wasting the perspective and insights only ownership can bring.

5. The Law of Focus

Priority focus is about the difference between activity and targeted results. Individuals know their metrics and are riveted on the area of the business where they can achieve the most direct impact on strategic goals. Strategic planning precedes decisions about use of company funds, facility use, the non-renewable resource of time.

6. The Law of Harvest

Every business has its seasons in which it must prepare (spring), nurture (summer), harvest (autumn), and pass on (winter). Its success in each season is directly correlated to the disciplines sown in the other three seasons. Although most of us recognize the Law of Harvest, which requires that ownership harvest in the right season, and not at times when the business should be preparing itself to bear fruit. But ownership must also understand the disciplines required and the most important priorities during Start-Up, Growth, Succession. By knowing at all times where the business is in its life cycle, ownership can provide for the needs of the business, anticipate the need to change and reinvent, and develop a capital plan—all at the right time.

7. The Law of Balance

For optimal performance, the business must be balanced in every dimension of its being—price and service, growth and equilibrium, innovation and standardization, speed and quality, cash flow and long-term investment. This Law of Balance also applies to ownership and leadership, who must be deliberate in their efforts to be healthy and balanced in their own lives. The business is a mirror of ownership's choices.

Each of these Laws have implications for our choices and decisions as owners. They require us to remember that our focus is to meet the needs of the business before we meet our own needs. Because when we are diligent about taking care of the business, the business will take care of us.

That isn't a law, but it's a strong belief I have that's proven over the years to be pretty reliable.

SEASONS
...of the Living Enterprise
Spring
Plant
Prepare
Summer
Nurture
Grow
Autumn
Harvest
Steward
Winter
Pass on
to the new

High-Performance Company

THE HIGH-PERFORMANCE COMPANY ACHIEVES PROFITABLE RESULTS BY WAY OF PRACTICES AND DISCIPLINES THAT ENABLE IT TO ACHIEVE A SIGNIFICANT LEVEL OF SUCCESS THROUGH AN ONGOING PROCESS OF THINKING, EXECUTING AND ACCOUNTING ON A CONSISTENT BASIS.

Chapter Six
Practices & Conditions of High-Performance Companies

"For optimal performance, the business must be balanced in every dimension of its being."

A Great Enterprise is inevitably a high-performance company. It executes well and it executes consistently. It leverages strategic thinking to develop successful tactics. It works from a Think-Execute-Account mindset.

The result that is always in the crosshairs of a high-performance company? Profitability.

The practices and conditions that support high performance show up in five key areas of the enterprise: Vision, People, Strategies, Systems, Beliefs and Values.

Vision

High-performance companies:

- Have a vision ***people believe*** in and identify with. They see themselves as a part of the vision and therefore it gives them a sense of belonging and connection. The vision creates a "we" mentality instead of a "them and us" mentality.
- Give people something to aspire to.
- Believe its vision will ***make a positive difference*** when it is achieved.

People

High-performance companies:

- Have a visionary leader who communicates freely, inspires trust, values people and their ideas, is focused on results and has the courage to make tough decisions.
- Grow and develop people as a precious resource that, when cultivated, will make the business more valuable. In other words, they ***treat talent like capital.*** They slot it well. They constantly examine skill sets, core competencies, capabilities and giftedness, and put the conditions in place that allow people to work out of their highest and best value.
- Has strategic thinking across disciplines through bold and honest communication.

- Are led by a team of executive leaders who are synergistic while encouraging divergent thought from individuals. Group think does not contribute to high performance.
- Interact respectfully.
- ***Empower leaders*** to make decisions at the level closest to problems, creating a sense of ownership and incorporating the insights of those who live with the problems.
- Have ***faith in one another*** tied to a belief in the good intentions and the spirit of others on the team. People know they're in an environment where they are trusted, so they ***feel free to innovate***, to generate and to grow.
- Inspire people to give of themselves beyond their role descriptions.

Strategies & Systems

High-performance companies:

- ***Tie compensation to results.*** Variable-based compensation is a basic assumption—when the company performs well, it's leaders are rewarded. The inverse also applies. When the company doesn't perform, there are consequences, not for the purpose of punishment, but as incentive.
- Outcomes and timeframes are clearly defined and progress is measured systematically.

- ***Organize around performance units*** of profit centers within which leaders are accountable for achieving outcomes that are within their control. These performance units are measured autonomously as well as interdependently, to encourage people to own the results they can affect and to pull together for the common good.
- Adjust quickly.
- Structure processes and systems around meeting the needs of customers and clients, not around meeting the needs of the people within the business.

> Every company sees the world through its own filters; this creates the particular world view that is the basis for its defining beliefs and values.

- ***Earn relationship trust***, both within the business and externally, with clients or customers, strategic allies, vendors, by following through on every commitment and every promise
- Know their numbers—net profit, gross profit, equity to debt, sales per month, profitability by business center, and on and on. ***Everyone knows what is necessary to win.***

Beliefs and Values

In addition to the principles surrounding strategy and talent and vision that contribute to the culture of a high-performance company, companies that build Business Wealth are invariably founded on certain beliefs and values.

Specific beliefs and values vary from business to business. And if ownership does not define and communicate those beliefs and values, they will evolve by default. Unspoken beliefs and values that evolve by default often don't contribute to the health and well-being of an enterprise.

Here's what to look for in a business operating from healthy beliefs and values:

- Leaders set a high-performance standard, ***modeling behaviors***, habits and practices.
- Mistakes are integral to continuous learning and view them as ***opportunities to adjust course***.
- People share common goals and commitments.
- They care about details. The small choices we make every day ***reflect what we really believe.***
- Problems are handled ***assertively but positively;*** it is recognized that people throughout the organization can help resolve problems.
- They follow ***prudent business practices***, both on and off the balance sheet.

Owners see the world through their own filters; this creates a company world view that is the basis for its defining beliefs and values.

Because the work we do with business owners brings us into deep relationship with owners and their leaders, I believe it's important for owners to understand our world view here at The Performance Group, the Performance Capital Group and the Institute for Business Owners. When we work with a business, we're asking people to trust us with their future.

Here are the beliefs we try to live by in order to earn the trust and confidence of business owners.

- We believe God owns everything. So it follows that we believe ***leaders are only stewards*** of God's resources. They are given the responsibility and accountability for managing those resources in direct proportion to the abundance with which they've been blessed.
- We believe there is ***power in knowledge*** and when power is provided to the right kinds of leaders, healthy enterprises will result. These healthy enterprises will result in a society made up of people who are economically healthy and thus better able to contribute to the health of their local, regional, national and global communities.
- We believe ***one leader at the top*** of every organization determines the culture of the enterprise and the opportunity for that enterprise to reach its full potential. This, too, is stewardship.
- We believe every individual has a ***God-given talent*** that, when fully tapped, has the potential to bear the fruit of genius.
- We believe the single greatest obstacle to releasing the asset potential of private enterprise is the unwillingness of ***ownership to adapt*** its perspective during the continuously-changing life stages of the business.

- We believe that to build a Great Enterprise, owners must make a ***conscious choice to serve the needs of the enterprise*** before serving their personal desires. When this trade-off is made, paradoxically, the rewards are greater personally, organizationally and economically.

Building a high-performance company doesn't occur by implementing certain policies or procedures. High-performance is itself a world view. It represents ongoing choices to be made by ownership, choices that must be authentic and integrated.

When those choices are made for the practices and conditions, the beliefs and values that are core to high-performance, a healthy organization results. And a healthy organization is capable of building the kind of Business Wealth that can lead to True Wealth.

Business Wealth

A state of significant economic well-being that results from consistently applying certain financial and strategic principles that enable a business to mature into an enterprise that serves as the source of balance sheet wealth for its owners and stakeholders.

Chapter 7
From Business Wealth to True Wealth

"The practices they've employed to achieve Business Wealth have created something they never imagined—True Wealth."

Most of our clients hear us talk a lot about Business Wealth™.

We help them implement principles and practices that lead to the kind of healthy business that is the basis for balance sheet wealth. We don't believe any business owner needs to be satisfied with keeping the doors open, making payroll, sweating on a treadmill indefinitely. So we show them a pathway to significant financial return that can be used outside the business to capitalize other opportunities.

At the root of this Business Wealth concept is the notion that business owners don't need to be hard-working employees of their businesses. Like any good steward, they can put their money to work outside the business, which positions them to be a better capitalist to the business.

In this way, the owner achieves balance sheet wealth and the business is well-capitalized to take advantage of opportunities for enterprise growth.

When we talk with owners about this Business Wealth System, they tend to think Business Wealth is the end-game. They have the resources to capitalize other opportunities. They can invest and diversify. They've accomplished the level of freedom that served as a spark for their dreams of ownership.

But as we all know, everything changes.

And the day will come when these owners realize that the practices they've employed to achieve Business Wealth have created something they never imagined—they now possess True Wealth.

These publications by Sam Frowine
are available
at
www.ThePerformanceGroupUSA.com

Lifeblood:
How Successful Business Owners Achieve Wealth

Blueprint
for Building Great Enterprise

Journey Map
A Journal for Business Owners

Ownership Wealth
A bi-weekly e-newsletter for business owners

Business Wealth, True Wealth and Great Enterprise are trademarks of the Center for Great Enterprise.

FOUNDATIONS for Great Enterprise and True Wealth. Copyright 2000, 2003 and 2007 by Sam Frowine. All rights reserved. Printed in the United States of America. No part of this book may be reproduced or transmitted in any form or by any means, electronic or mechanical, including photocopying, recording, or by any information storage and retrieval system, without written permission from the publisher. For information, address the Institute for Business Owners, 9115 Harris Corners Parkway, Suite 330, Charlotte, NC 28269.

Made in the USA
Monee, IL
04 November 2026

67374652R00039